Tutu and Zazu
Make Chocolate

Tamar Laderman

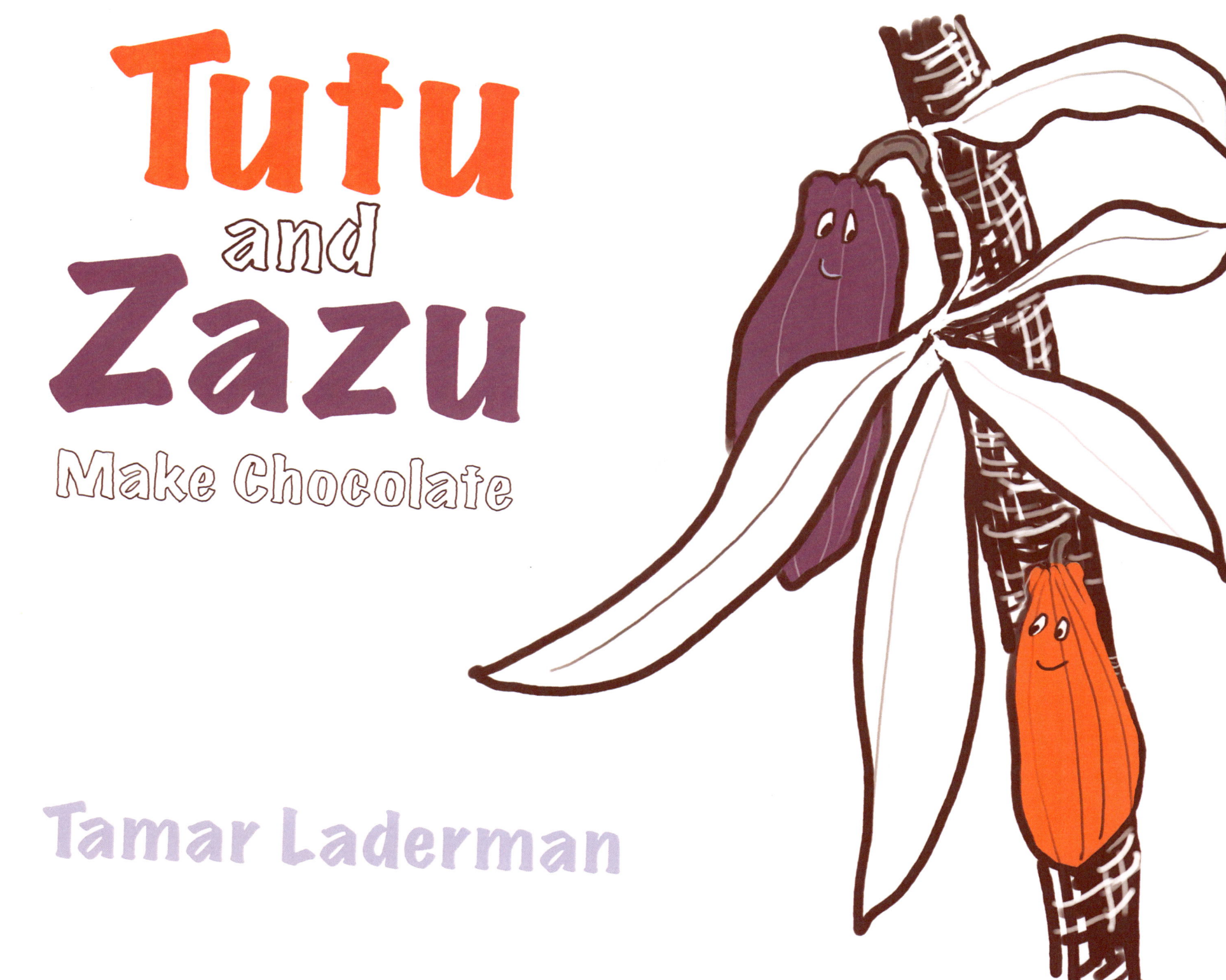

Copyright © 2025 Tamar Laderman
Edited by Donna Racik

All rights reserved. No part of this publication may be reproduced, distributed or transmitted in any form or by any means, or stored in a database or retrieval system, without the prior written permission of the copyright holder.

ISBN 979-8-9927346-0-7 Hardcover

First Edition

For Chocolate Lovers Everywhere
(Especialy for Carol, Ossia, Ezra, Isabel, Ayla, Emma, Page & Olivia)

Tutu and Zazu love the Big Island of Hawaii.
They love the earth and the sea.
They love their family and friends.
They love each other,
and best of all they both really love chocolate.

There are oceans and volcanoes
and the earth is blessed with rain and sun.
Tutu and Zazu work and play with family
and friends
and best of all they make chocolate.

Tutu and Zazu shine with
pleasure.
The trees love their new
friends.
They grow tall.
The trees flower
and make many many yellow
and red cacao pods.
Beautiful!

Tutu and Zazu are so excited.
They are ready to harvest the pods.
When the buckets are full,
the cracking begins.

The pods are very hard,
and the yummy cacao seeds are nestled snugly inside.
Zazu uses his special cracker to open the pods.
Crunch Crunch
CRACK!

Look what's inside!
Here they are!
The cacao beans that
make the chocolate!
Tutu and Zazu can smell
the sweet fruit.
They scoop out the beans
like ice cream.
Yum!

Tutu and Zazu wrap the sweet cacao
beans in banana leaves.
Bubble, fizzle, and gurgle.
The cacao beans are fermenting.
Burp!

They are so slippery.

Time to get dry!
Tutu and Zazu carefully spread
the cacao beans on racks.
It's warm in the sun.
Ahhh.

What flavor is this batch?
It's time to taste.
Sometimes the cacao beans are fruity.
Sometimes the cacao beans are nutty.
Sometimes the cacao beans are spicy.
Tutu and Zazu love them all!

Tutu pours cacao beans into the roaster.
The beans get very hot.
Zazu takes a deep breath.
Ooh…
Smell the chocolate!

Crank crank crank.
More cracking.
The beans are separated into
nibs and husks.
They rattle down the slide.
Wheee!

Whoosh!
The husks are blown away.
Tutu and Zazu giggle.
Look at all the nibs!

Whew!!
Almost there....

Whirr, whirr, hum, hum.
The grinding stones grind the nibs.
They are silky smooth.

Woohoo!!
It's like magic!
Tutu and Zazu jump for joy!

Tutu and Zazu taste and sniff.
They drizzle in honey
and sprinkle in nuts.
Delicious!
Sweet yummy chocolate!

It's time to share!
Hooray!!
The family and friends dance and sing together.
Yum Yum Yum!!!!

100% HAWAII
SMALL BATCH CRAFT CHOCOLATE

HONEY-SWEETENED CHOCOLATE
with Cacao Nibs

HONEY-SWEETENED CHOCOLATE

SUPER DARK CHOCOLATE
with Cacao Nibs

HONEY-SWEETENED CHOCOLATE
Hawaiian Hot Chile Pepper & Kona Sea Salt • 80% Dark

HONEY-SWEETENED CHOCOLATE
with Coffee

HONEY-SWEETENED CHOCOLATE
with Macadamia Nuts & Sea Salt

HAKALAU CHOCOLATE

$8/bar

100% Big Island | Honey-Sweetened Chocolate

Tutu and Zazu love the Big Island of Hawaii.
They love the earth and the sea.
They love their family and friends.
Tutu and Zazu love chocolate and best of all,
they love you!

www.ingramcontent.com/pod-product-compliance
Lightning Source LLC
Chambersburg PA
CBRC102342090526
44582CB00015B/193